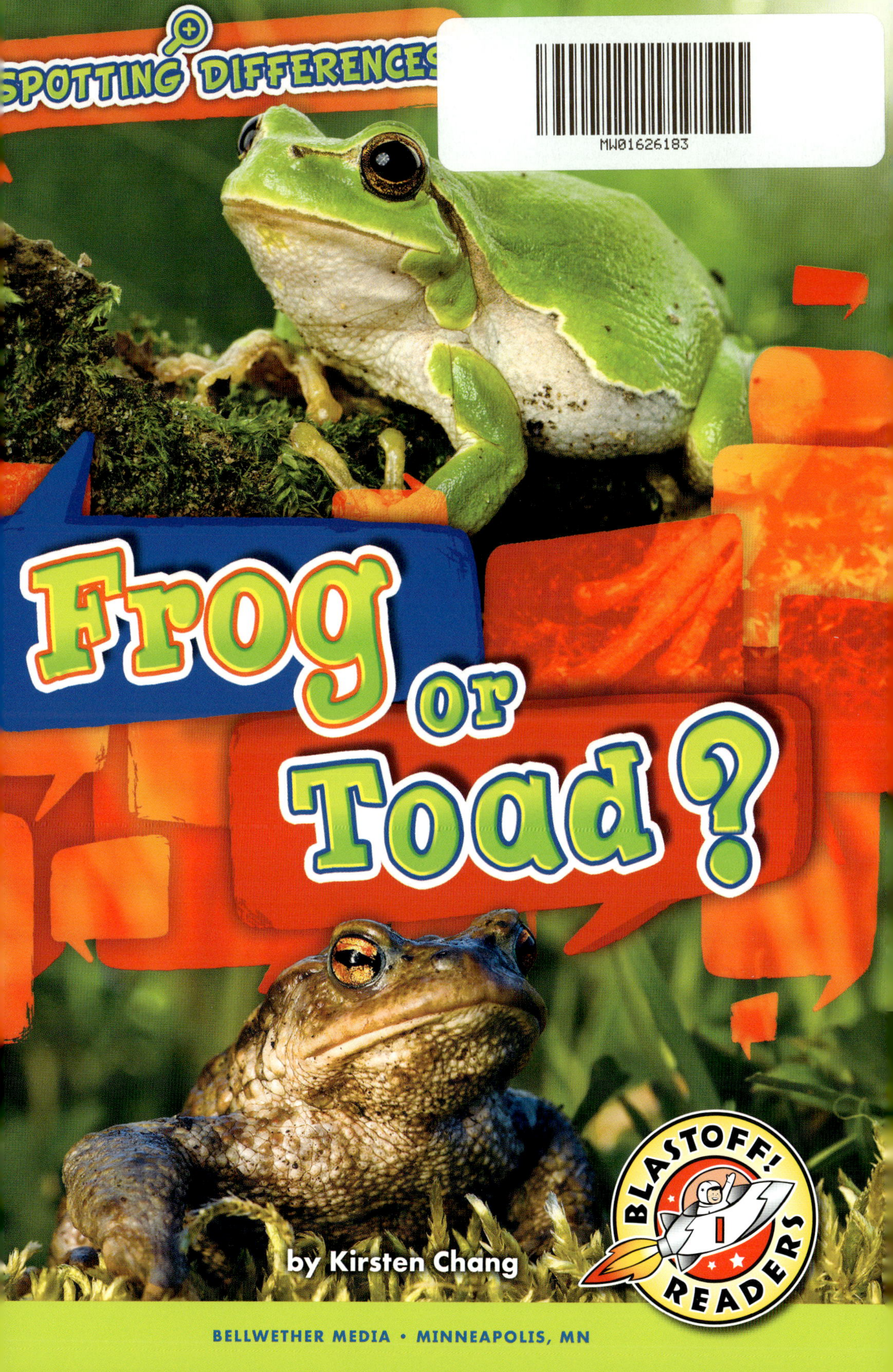
SPOTTING DIFFERENCES
MW01626183
Frog or Toad?
by Kirsten Chang
BLASTOFF! READERS
1
BELLWETHER MEDIA • MINNEAPOLIS, MN

Blastoff! Readers are carefully developed by literacy experts to build reading stamina and move students toward fluency by combining standards-based content with developmentally appropriate text.

Level 1 provides the most support through repetition of high-frequency words, light text, predictable sentence patterns, and strong visual support.

Level 2 offers early readers a bit more challenge through varied sentences, increased text load, and text-supportive special features.

Level 3 advances early-fluent readers toward fluency through increased text load, less reliance on photos, advancing concepts, longer sentences, and more complex special features.

★ **Blastoff! Universe**

Reading Level

Grade K

Grades 1–3

Grade 4

This edition first published in 2021 by Bellwether Media, Inc.

Library of Congress Cataloging-in-Publication Data

Names: Chang, Kirsten, 1991- author.
Title: Frog or toad? / by Kirsten Chang.
Description: Minneapolis, MN : Bellwether Media, Inc., [2021] | Series: Blastoff! readers: spotting differences | Includes bibliographical references and index. | Audience: Ages 5-8 | Audience: Grades K-1 | Summary: "Developed by literacy experts for students in kindergarten through grade three, this book introduces frogs and toads to young readers through leveled text and related photos"-- Provided by publisher.
Identifiers: LCCN 2019054183 (print) | LCCN 2019054184 (ebook) | ISBN 9781644871980 (library binding) | ISBN 9781681038223 (paperback) | ISBN 9781618919564 (ebook)
Subjects: LCSH: Frogs--Juvenile literature. | Toads--Juvenile literature.
Classification: LCC QL668.E2 C424 2021 (print) | LCC QL668.E2 (ebook) | DDC 597.8/9--dc23
LC record available at https://lccn.loc.gov/2019054183
LC ebook record available at https://lccn.loc.gov/2019054184

Editor: Elizabeth Neuenfeldt Designer: Jeffrey Kollock

Printed in the United States of America, North Mankato, MN.

Table of Contents

Frogs and Toads

Frogs and toads are **amphibians**. They can live in water and on land!

frog

Frogs and toads catch **prey** with long, sticky tongues. How are they different?

toad
tongue

Different Looks

Toads usually have round bodies. Frogs are often **narrow**.

Frogs have long back legs. Toads have shorter back legs.

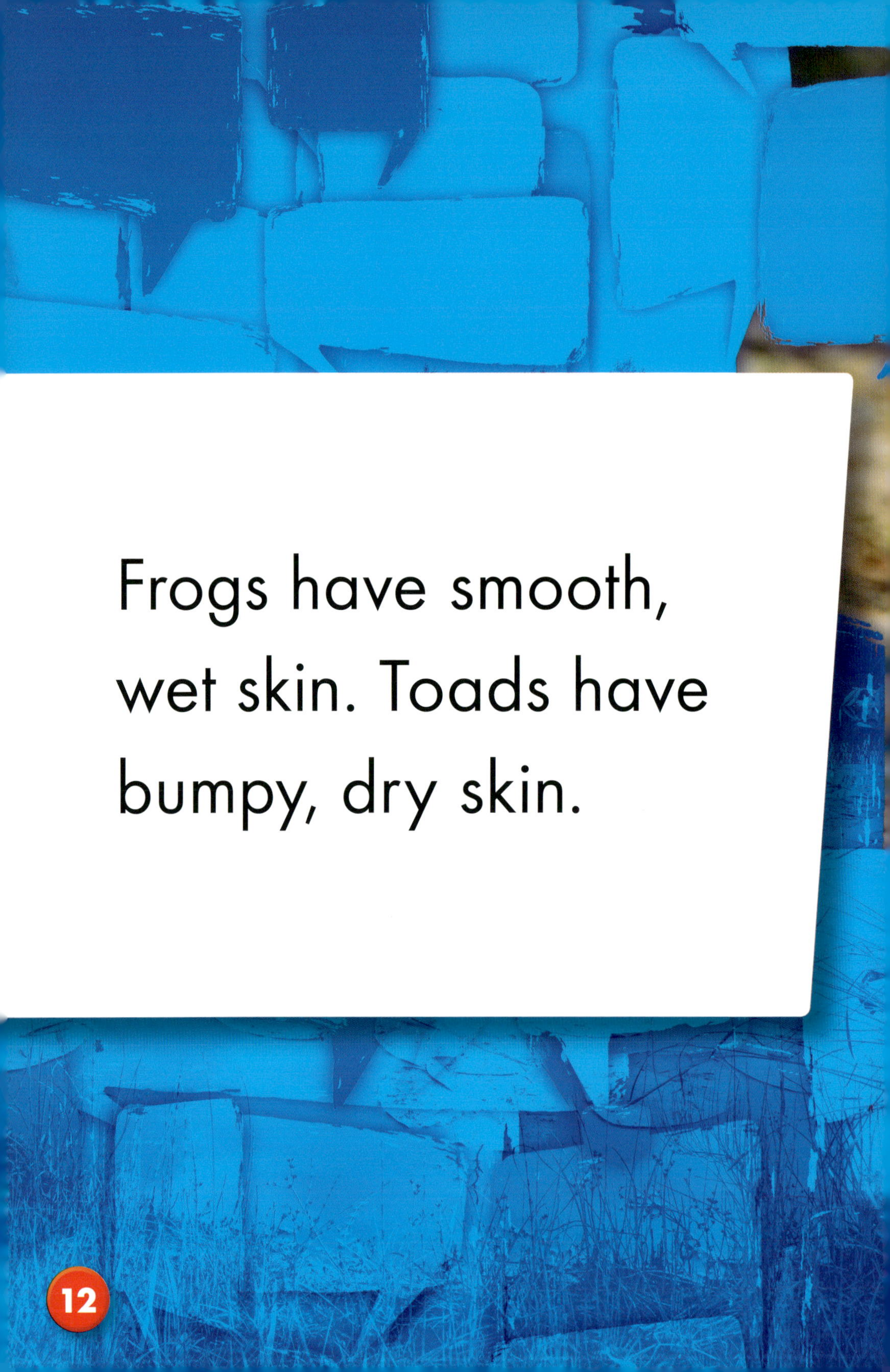

Frogs have smooth, wet skin. Toads have bumpy, dry skin.

bumpy
skin

Different Lives

Frogs can live in water, on land, or in trees. Toads mostly live on land.

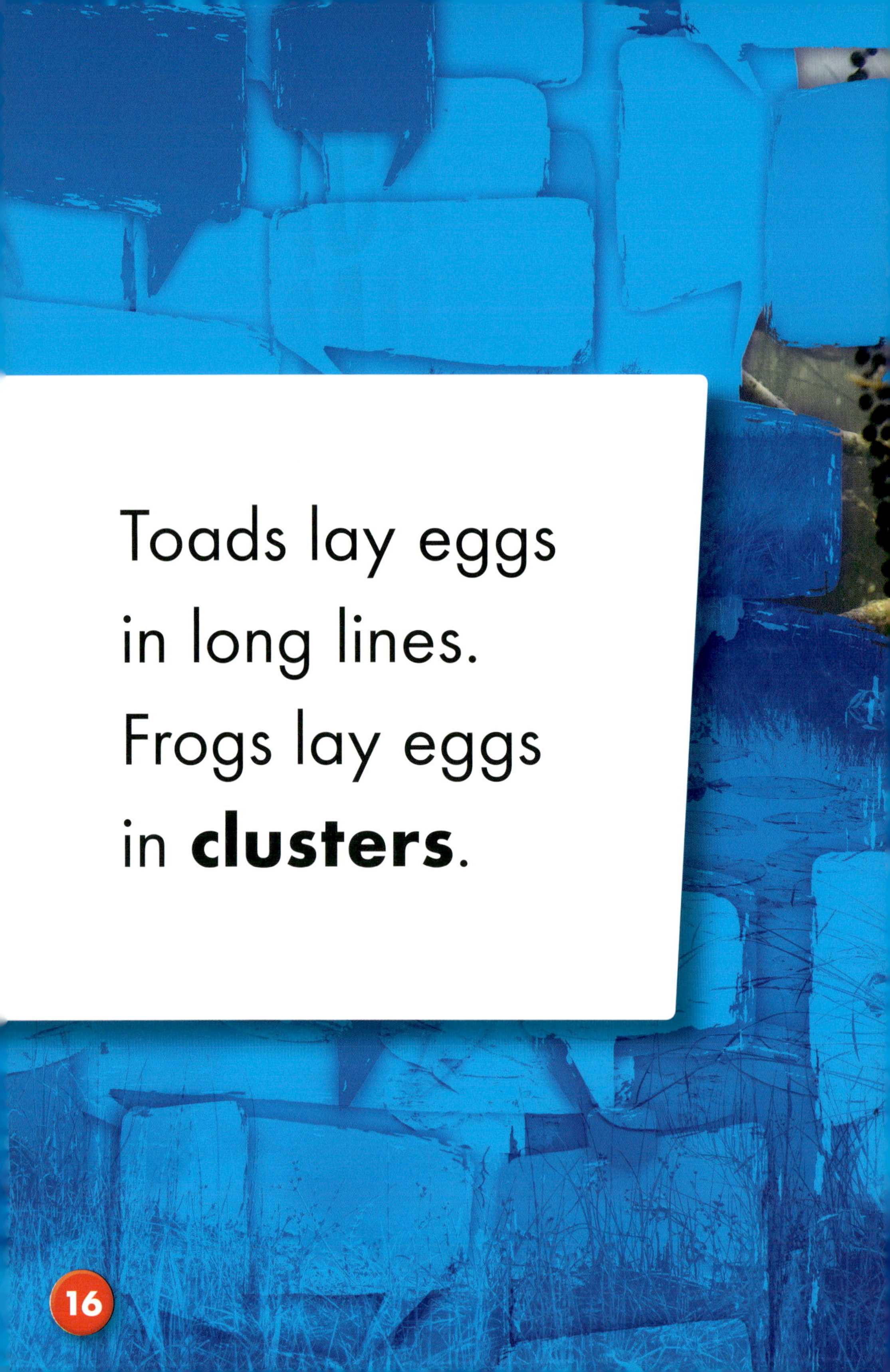

Toads lay eggs in long lines. Frogs lay eggs in **clusters**.

toad
eggs
cluster

Frogs can jump very far.
Toads hop or walk.
Which one is this?

Side by Side
narrow body
long back legs
smooth, wet skin
Frog Differences
can jump far
live in water, on land, or in trees
lay eggs in clusters

round body
short back legs
bumpy, dry skin
Toad Differences
hop or walk
mostly live on land
lay eggs in lines

Glossary

amphibians

animals that can live on land and in water

narrow

thin

clusters

similar things that are grouped closely together

prey

animals hunted by other animals for food

To Learn More

AT THE LIBRARY

Jenkins, Steve, and Robin Page. *The Frog Book*. Boston, Mass.: Houghton Mifflin Harcourt, 2019.

Kralovansky, Susan. *Frog or Toad?*. Minneapolis, Minn.: ABDO Publishing, 2015.

Murray, Julie. *Amphibians*. Minneapolis, Minn.: Abdo Zoom, 2019.

ON THE WEB

FACTSURFER

Factsurfer.com gives you a safe, fun way to find more information.

1. Go to www.factsurfer.com.

2. Enter "frog or toad" into the search box and click 🔍.

3. Select your book cover to see a list of related content.

Index

The images in this book are reproduced through the courtesy of: Manuel Findeis, front cover; Pavel Krasensky, front cover; Svetlana Foote, pp. 4-5; Bildagentur Zoonar GmbH, pp. 6-7; davemhuntphotography, pp. 8-9; Pascale Gueret, pp. 10-11; JaklZdenek, pp. 12-13; Alex Stemmer, p. 13 (bubble); gravinerva, pp. 14-15; BIOSPHOTO/ Alamy, pp.16-17; Joanna K-V, p. 17 (bubble); Ernst Dirksen/ Buiten-beeld/ SuperStock, pp. 18-19; Fablok, p. 20 (frog); kyslynskahal, p. 20 (left); MladenR79, p. 20 (middle); Clara Bastian, p. 20 (right); Patrick K. Campbell, p. 21 (toad); Wouter Marck, p. 21 (left); Phuketian.S, p. 21 (middle); Huaykwang, p. 21 (right); Kontrastwerk, p. 22 (amphibian); Ellita, p. 22 (clusters); Vaclav Sebek, p. 22 (narrow); Louis.Roth, p. 22 (prey).

Our Irish Heritage

A Kid's Guide to the Irish Diaspora

This book belongs to:

____

Abair Linn Publishing

Written by Rachel J. Cooper
Illustrated by Lu Ó'Mochóir
Design and Layout by ruthgunning.com

Who are the Irish diaspora?

Well, first of all it's a hard word to pronounce, isn't it! It's pronounced "die-as-pour-ah".

A diaspora is a word that is used to describe a large group of people who have moved over many years from the country they were born in, out into other parts of the world.

The Irish diaspora are the families who moved from Ireland to other countries around the world. Most of them traveled many miles across the Atlantic ocean to the USA. This book will teach you all about Irish heritage, language, history, and culture. Are you excited to learn? Then let's begin!

In Ireland, like many other countries long ago, there were some sad times in its history, and sometimes there wasn't much money for everybody to live comfortably.

Families were a lot bigger back then; sometimes there would be as many as ten or more children! That's a lot of boys and girls to feed and keep warm and clothed. So, it made sense for many years for people to leave Ireland to start better lives in places such as America, the United Kingdom, and Australia.

To start their new lives, Irish children and their parents boarded big ships together and set sail from the shores of Ireland.

You might have a great, great, great, great (that's a lot of greats!) grandfather or grandmother who originally came from Ireland and traveled on one of these large ships.

They never got to see Ireland again, or the people that they left behind, which was very sad for them. This is why it is so important for all the boys and girls born into the Irish diaspora in America, and the rest of the world, to know where your ancestors came from so that you can keep your Irish heritage and identity alive in your hearts, and make your ancestors very proud indeed.

Irish people continued to leave Ireland throughout the years, so you might also be a more recent generation of the diaspora, whereby you could have a parent or a grandparent who arrived by plane from Ireland. You might get to visit your Irish relatives or they come to visit you.

No matter how long ago or how recently your relations arrived from Ireland, being part of the Irish diaspora is something to be very proud of and something that you should always remember.

If you have what's called "Irish Heritage" - perhaps you might like to learn a bit about Ireland and your heritage - which is the word for your family's past culture.

Irish Surnames

You might have a surname that is Irish - like O'Connor, Sullivan, Butler, Fitzgerald, O'Reilly, Walsh, Doyle, Ryan, Kennedy, Kelly, O'Brien, Murphy, MacCarthy, Gallagher, Byrne, O'Neill, Lynch, Quinn, Moore, O'Connell, Burke, Brennan, Maguire, Flynn, Nolan, Duffy, O'Mahony, Healy, O'Shea, or Doherty. There are so many Irish Names it would be impossible to list them all. If yours is not mentioned above be sure to write it at the beginning of the book.

'O' before a name means that you're a descendant of the original person of that name. So, if you're an O'Sullivan, your family are from the O'Sullivan clan (clan means family). If you're a Mc or Mac it means "the son of". The names were anglicized or changed into English from the original Irish.

Many names were also changed when people moved to their new country. Sometimes, names were hard to understand and were changed when people arrived from Ireland.

You might have heard of the city, town, or village associated with your heritage!

Often, family stories were passed down through the years and you might even know where your family is from in Ireland. Ireland has some beautiful places, towns and villages, and, someday, you may even go to Ireland to find that place that you come from. You might even feel at home in that place!

p
q

The Druids and Celts

Ireland is a very ancient country and it was the home to many different people - from Celts and Druids to Vikings, and Normans. Let's talk a little bit about them.

Celts and Druids lived together in Ireland over 2,000 years ago. The Celts were very musical and advanced for their time. They also loved to dance. They made all their hunting weapons from iron and they made jewelry, art, and crafts. They farmed the land and worshipped many Gods.

The Druids were the priests and elders of the Celts. The Celts would go to the Druids to ask for advice on how to keep their Gods happy, so they would have plentiful food from the land, and that their families were protected from harm.

If any little Celtic boys or girls got sick, the Druids would give them medicine made from plants. They were very knowledgeable. The Celts were firm believers in ghosts and omens so they would ask the Druids for help from the gods to protect them from evil spirits.

The Vikings

The Vikings were fierce warriors who arrived in Ireland from Scandinavia to live amongst the Irish after the Celts and Druids. Scandinavia is a place in North Europe. You might want to ask a teacher, parent, or guardian to show you on the map where that is.

These Viking warriors sailed from across the sea and arrived on Irish shores in boats called Longships. Viking children settled in Ireland with their parents. The children mostly helped out farming the land and preparing food, but they also were good with woodwork and helped to build the Longships. Some of Ireland's cities including Dublin and Waterford were named by the Vikings.

The Normans

The Normans arrived on Irish shores from England after the Vikings in 1169. It was the Normans who introduced the English language to Ireland.

English is still the spoken language of Ireland today, just like in the USA, but Irish is the national language and all Irish children learn it in school to this day. The Normans built stone castles to live in Ireland, which also provided protection from danger. The Irish clans were not very happy that the Normans took their land and tried many times to get it back. These castles stood against the invasions of the Irish clans. The children in these castles were taught how to use bows and arrows and played with wooden swords.

You can still see Norman castles all over Ireland today. Towns like Carlow, Trim, and Kilkenny were all built around their Norman castle.

Halloween

Ireland is where Halloween came from! Did you know that? It was called Samhain (pronounced Sow-in) by the ancient Celts of Ireland, who we mentioned earlier. They began to celebrate it over 2,000 years ago.

The ancient Irish Celts first celebrated Samhain as a time of gatherings and feasts in October to prepare for the harsh winter ahead. They believed that the spirits or fairies could find their way into our world at the time of Samhain. They lit big bonfires to ward away evil spirits.

People would light Jack O'Lanterns, set tables, and dress in costumes with extra places for the good spirits of their relatives, who would help bring about less harsh winter. Sound familiar? This tradition was brought by Irish people to America and that's how Halloween became a celebration, just like it was for many centuries before in ancient Ireland. So, as part of the Irish Diaspora you should be very proud that this popular time of year originated in Ireland.

Irish dancing and traditional Irish music

Many of the Irish diaspora boys and girls learn Irish dancing, but do you know where it came from and why it is still so popular today?

It is said that the Irish Celts (who we mentioned a bit about before) were responsible for introducing dancing to Ireland. Irish dancing as we know it today was developed through the years. This popular dance is known for its tap-like rhythms. The steps are danced in two forms. The first is tapped very quickly in hard shoes that make a loud bang on wooden floors. The second is danced in soft ballet-like shoes.

Irish dancing costumes are bright and colorful and would have originally been made from Irish tweeds, laces, and linens. The embroidery designs down the front were taken from books with celtic designs.

Sometimes, the girls curl their hair so their ringlets bounce in rhythm along to the music. Irish traditional music plays a very important part in Irish dancing. You will hear fiddles, the accordion, the tin whistle, the uilleann pipes (pronounced ill-in pipes), and the bodhrán drum (pronounced bow-rawn) all being played to get a good rhythm going.

We cannot mention Irish dancing and the Irish Diaspora without mentioning Michael Flatley. He was responsible for re-introducing Irish dancing to a global audience through his show Riverdance, when he originally danced in Ireland with Jean Butler back in 1994.

His more modern version of Irish dancing introduced the use of the arms and different stepwork, compared to the traditional version of Irish dancing where dancers have their arms straight down by their side.

Every year, many people who have Irish heritage or love Irish dancing participate in classes and competitions around the world keeping this ancient art form alive!

The GAA

The GAA or Gaelic Athletics Association was founded way back in 1884 in county Tipperary. It is the official organisation which promotes Irish sports such as hurling, Gaelic football, camogie, handball, and rounders. The children of the Irish diaspora should know a little bit about it, as it is an extremely important part of Irish culture both in Ireland and in the USA.

Gaelic football and hurling are the most popular sports promoted by the GAA. Let's learn a little bit about them.

Hurling is a stick and ball game. The stick is called the hurley stick and players use it to catch and hit the ball (called the sliotar) between the goalposts.

Gaelic football is a little bit like soccer but in this game the players can catch and punch the ball as well as kick it into the net.

Did you know that there are over 130 GAA clubs in the USA? Ask your parents/guardian where the nearest club is to your house, maybe they might take you there one day!

Saint Patrick

Saint Patrick is the patron saint of Ireland. Saint Patrick's Day is on the 17th of March every year. Parades are held all over the world, to celebrate Irish culture. Does your family celebrate this special day? Saint Patrick was born in Britain over two thousand years ago. When he was 16 years-old he was kidnapped by pirates who took him to Ireland. He escaped, but he returned to Ireland to teach the people about God. Legend has it that Saint Patrick drove the snakes out of Ireland!

The Harp

Many centuries ago it was reported around the world that the Irish were excellent harp players, which is why Ireland adopted the harp as an emblem. Ancient harpists were highly prized and considered the pop stars of their day.

The Shamrock

The Shamrock became known throughout the world as Ireland's National Emblem. Saint Patrick used it to teach his pagan followers who were converting to Christianity the Holy Trinity of God, Jesus Christ and the Holy Spirit. Today, the Shamrock is recognised worldwide as Irish and loved! Who would have thought such a little shoot or sprig would be so well-liked!

Saint Brigid

Saint Brigid is Ireland's female patron saint. Her father was a pagan chieftain of Leinster and her mother was a Christian. Saint Patrick inspired her to deepen her faith and spread the word of God. Her feast day is on the first day of February and it is now a holiday in Ireland. Saint Brigid's Day occurs at the same time as Imbolc, the ancient pagan festival of spring.

So now we are going to learn a little bit about the Irish national anthem, which is called Amhrán na bhFiann (ow-ran na vee-un), and you can see all the words on the next page.

The Irish national anthem was written by two friends: Peadar Kearney and Patrick Heeney, back in 1909-10. They were from Dublin city, which is the capital city of Ireland.

It was first written in English as The Soldier's song but it was then translated into Irish or Gaelic by Liam O'Rinn and renamed Amhrán na bhFiann. Liam was also from Dublin and grew up very close by to Peadar and Patrick. The Irish national anthem was officially adopted by the State in 1926, and Irish people have been singing it ever since.

The first line of the Irish anthem says that we are all soldiers. And this is still true. Today we all have to fight together against things like inequality, bullying and unfairness. We fight against things that we know are wrong.

The anthem also talks about people "from a land beyond the wave". This can be seen as a tie to people like you and your family who left Ireland over many years for countries like America. The people back in Ireland always remember you in their hearts.

The Irish National Anthem

Sinne Fianna Fáil,
Sheena Feena Fall

atá faoi gheall ag Éirinn,
Ataw fwee ghyall egg Ey-rin

Buíon dár slua
Bween dar sloo-ah

thar toinn do ráinig chugainn,
Har-teen duh rawnig hoo-ing

Faoi mhóid bheith saor
Fwee voy-age vess air

Seantír ár sinsear feasta,
Shanteer air shin-shur faa-sta

Ní fhágfar faoin tíorán ná faoin tráill.
Knee awg-far fween teer-awn naw fween traw-il

Anocht a théam sa bhearna bhaoil,
Ah-nucked ah hame sa varna ba-wail

Le gean ar Ghaeil, chun báis nó saoil,
Le gan ar g-wail cun bawsh noh sale

Le gunna-scréach faoi lámhach na bpiléar,
Le gunna skrake fwee lawve-ock na bill-air

Seo libh canaig' amhrán na bhFiann.
Shuh live con-eeg ow-rawn nah vee-un

Below is the English translation which talks of Ireland's history and proud people.

"The Soldier's Song"

Soldiers are we,

whose lives are pledged to Ireland,

Some have come

from a land beyond the wave,

Sworn to be free,

no more our ancient sire-land,

Shall shelter the despot or the slave.

Tonight we man the "bearna bhaoil"

In Erin's cause, come woe or weal,

'Mid cannons' roar and rifles' peal,

We'll chant a soldier's song.

So now we are going to learn a little bit about the Irish flag which is called the Tricolour. The flag is called the Tricolour because of its three colours of green, white and orange. The three colours represent a unification of everybody living in Ireland.

In the box below draw the American Stars and Stripes flag so that it can be seen together with the Irish flag on the opposite page. If you know of any other flags be sure to include them too.

In the map opposite put an X where the county is that your family came from and write its name in the below space.

Derry
ANTRIM
NORTHERN
IRELAND
Donegal
ATLANTIC OCEAN
Tyrone
Down
Fermanagh
Armagh
Sligo
Monaghan
Leitrim
Cavan
Mayo
Louth
Roscommon
Longford
Meath
IRISH SEA
Westmeath
DUBLIN
Offaly
Kildare
Galway
Laois
Wicklow
Clare
REPUBLIC OF
IRELAND
Carlow
Tipperary
Kilkenny
Limerick
Wexford
Waterford
Kerry
CORK
CELTIC SEA

Now you know about your Irish heritage, why not find out a little bit more from your parents, grandparents, uncles, or aunts?

Slán leat!
That means "goodbye for now" in Irish (Gaelic)

What were the names of your ancestors who first came to the USA? Write them here:

When did they arrive? Write the year in this box:

Made in the USA
Middletown, DE
11 March 2023